101 Catalysis
With Transition Metal
Compounds
Questions and Answers

Christoph Sontag, PhD

i

DEDICATION

This work is dedicated to my PhD father, Prof.Dr. Heinz Berke. Through his teaching I was introduced to the exciting world of transition metal chemistry.

PREFACE

If you are one of those people who wonder about strange chemical reactions taking place with transition metal compounds, then you are at the right place - this booklet gives you a brief introduction to the four elemental steps in catalysis. Mostly all catalytic cycles can be broken down into these basic steps, leading to a deeper understanding of important reactions used in industrial scale.
Please try to answer the question part first before moving to the answer part in order to check your comprehension.
In case you detect bugs or have suggestions for improvement, please let the author know: c.sontag@web.de

All drawings were done with the freeware version of Chemsketch 2015 and the mind maps using the freeware XMind6. Sources from the web are indicated together with their citations.
For an in-depth and extended study of the topics in this booklet, you may refer to the opensource document:
http://webdelprofesor.ula.ve/ciencias/isolda/libros/organometalicos.pdf

Now I wish you an interesting journey into the world of organometal catalysis.

Dr.Christoph Sontag
Phayao, Thailand, April 2016

CONTENTS

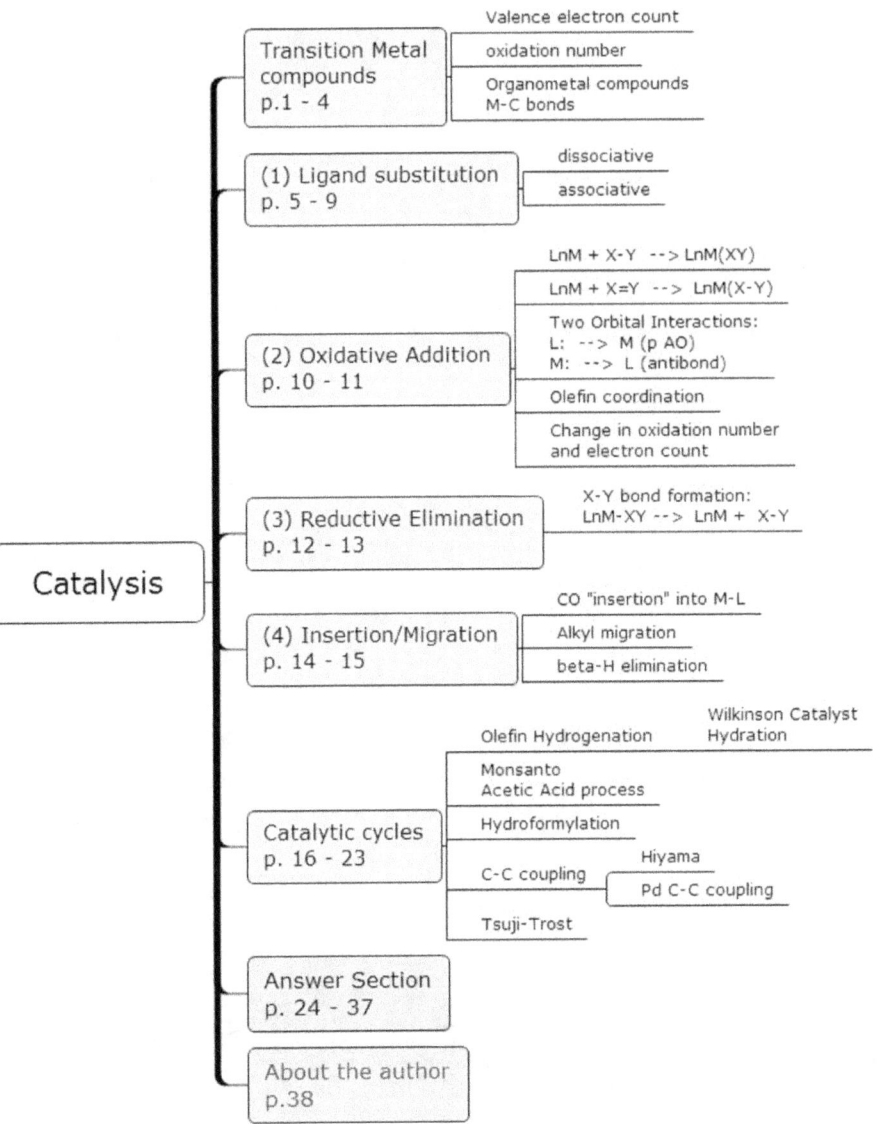

TRANSITION METAL COMPOUNDS

Depending on the position in the periodic table, we can see different behaviors of transition metals toward complex formation: (el = electron, TM = transition metal)

Valence electrons:

3	4	5	6	7	8	9	10	11
d^3	d^4	d^5	d^6	d^7	d^8	d^9	d^{10}	$d^{10}s^1$
21	22	23	24	25	26	27	28	29
Sc	**Ti**	**V**	**Cr**	**Mn**	**Fe**	**Co**	**Ni**	**Cu**
Scandium	Titanium	Vanadium	Chromium	Manganese	Iron	Cobalt	Nickel	Copper
39	40	41	42	43	44	45	46	47
Y	**Zr**	**Nb**	**Mo**	**Tc**	**Ru**	**Rh**	**Pd**	Ag
Yttrium	Zirconium	Niobium	Molybdenum	Technetium	Ruthenium	Rhodium	Palladium	Silver
57	72	73	74	75	76	77	78	79
La	**Hf**	**Ta**	**W**	**Re**	**Os**	**Ir**	**Pt**	Au
Lanthanum	Hafnium	Tantalum	Tungsten	Rhenium	Osmium	Iridium	Platinum	Gold

(see: http://www.chem.mun.ca/homes/cmkhome/
2011Notes_Chemistry_3211_Pt6_Organometallics.pdf

VALENCE ELECTRON COUNT IN TRANSITION METAL COMPOUNDS

In order to understand the bonding in a metal compound, we must be able to identify and count the valence electrons in such molecules (for an in-depth overview about ligands see: http://www.udel.edu/chem/dawatson/classes/Chem652_S13/lecture_notes/ 1b_structure_ligands_bonding.pdf

and: http://web.iitd.ac.in/~sdeep/Elias_Inorg_lec_5.pdf)

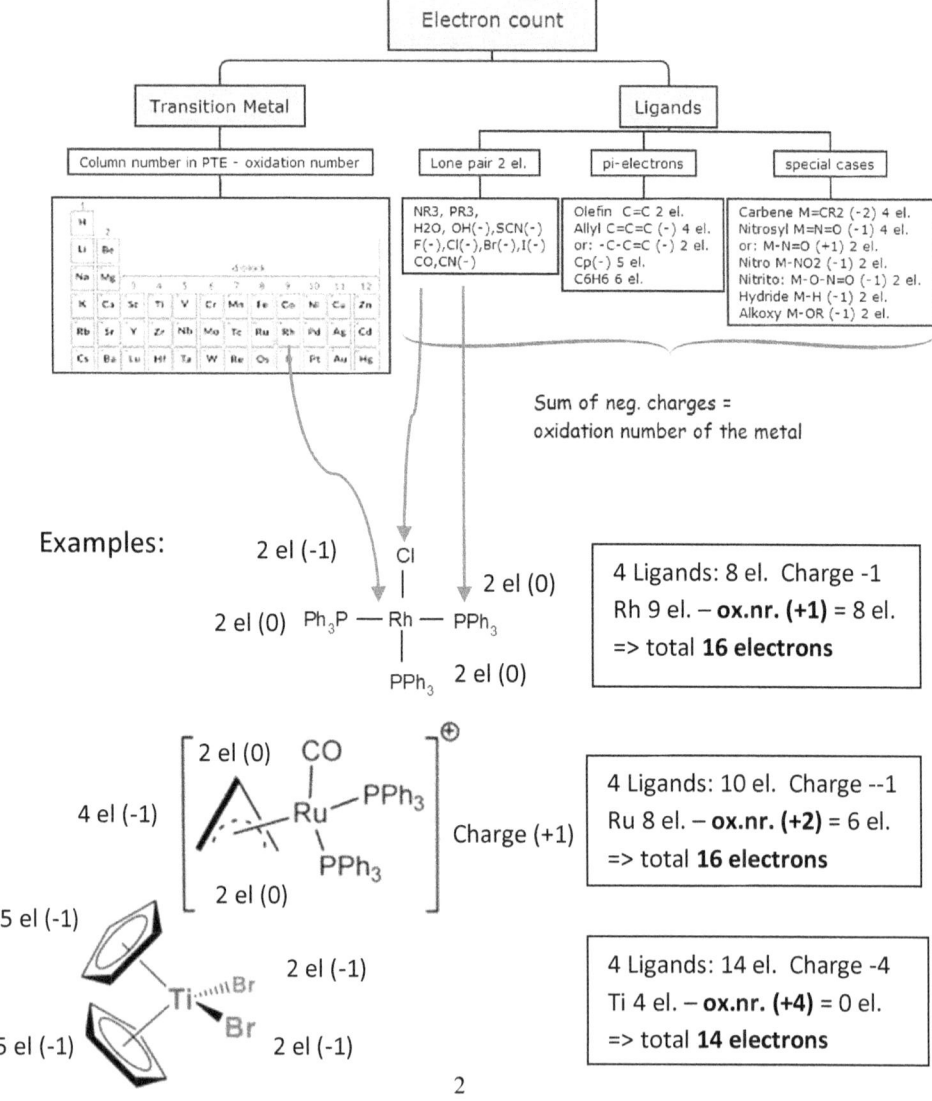

Find the oxidation number of the metal and the number of valence electrons in the following compounds:

| 1A | 1B | 1C | 1D |

Why is the oxidation number and the electron count important ?

It may seem that to find the metal oxidation number is just an academic task – but in fact it tells us a lot about the characteristics of the whole molecule. A higher oxidation number for example indicates low electron density on the metal. The electron count tells us if a complex is "saturated" or "satisfied" by its valence electrons.

See page 3: for early and late transition metals, a valence electron count of 14 or 16 seems perfectly ok, but a 16 electron complex of a middle transition metal tells us that it lacks two electrons and is actually a kind of underlined electrophile.

Which of the following compounds would be "satisfied" with their electron configuration ?

| 2A | 2B | 2C |

THREE KINDS OF TRANSITION METAL COMPOUNDS

In this compendium we are looking only at organometal compounds – because catalysis is all about forming new bonds (or breaking bonds) in organic molecules.

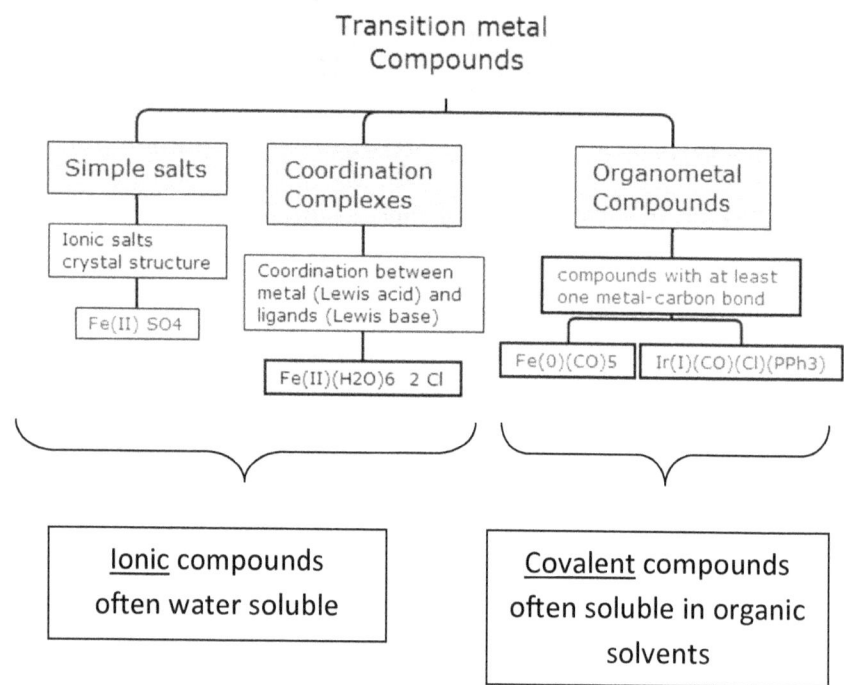

Catalytic Steps

Ligand substitution	dissociative: $$L_nM\text{-}X \longrightarrow L_nM \longrightarrow L_nM\text{-}Y$$ $-X$ $+Y$ associative: $$L_nM \longrightarrow L_nM\text{-}X \longrightarrow L_{n-1}M\text{-}X$$ $+X$ $-L$
Oxidative Addition	Splitting of Ligand bonds like H-H, C-H, C-C, C-X $$L_nM + \quad \overset{A=B}{\underset{A-B}{\longrightarrow}} \quad L_nM\overset{A}{\underset{B}{\diagup}} \qquad L_nM\overset{A}{\underset{B}{\diagup}}$$ $A - B$ $[L_nM\ A](+)\ B(-)$ Coordination of * Hydrogen * Olefins * R-X molecules
Reductive Elimination	Reverse reaction of oxidative Addition: Bond formation in the ligand and release of the ligand
Insertion / Migration	CO Insertion into M-C => acyl formation Olefin insertion into M-H => hydrogenation Alkyl migration => formation of C-C bond

(1) Ligand Substitution

The first basic step in catalysis is a ligand dissociation and replacement by another (normally stronger base) ligand:

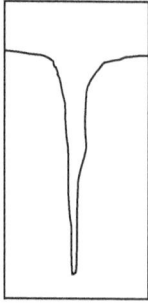
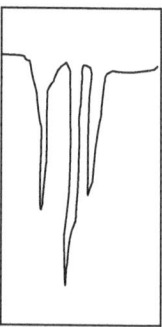

A common method to achieve this in the laboratory is UV-radiation of a carbonyl complex in a polar solvent such as acetonitrile CH_3CN or THF.

- UV Radiation causes loss of carbonyl in form of carbon monoxide gas, which is replaced by a nitrogen stream and therefore no longer available.
- The solvent molecules serve as temporary replacement of carbonyl even they are quite "weak" ligands (low basicity).
- Addition of another ligand molecule such as phosphine PR_3 will replace the solvent molecule in the complex.

The course of the carbonyl loss can be monitored by IR spectroscopy, which will show the number of remaining carbonyl ligands according to the typical pattern in CO-vibrations.

For $M(CO)_6$ as well as for trans-$M(CO)_4L_2$ we should see only one CO stretching peak: (O_h and D_{4h} symmetry)

For $M(CO)_5L$ as well as for cis-$M(CO)_4L_2$ we should see 3 different CO stretching peaks: (C_{4v} and C_{2v} symmetry)

Use of group-theory:

will be covered in another 101 series (Group theory for Chemists). Just a short comment here: IR vibration signals from carbonyls are very strong and the pattern can help us to identify the coordination. For a cis-$ML_4(CO)_2$ complex for example, the symmetry group would be:

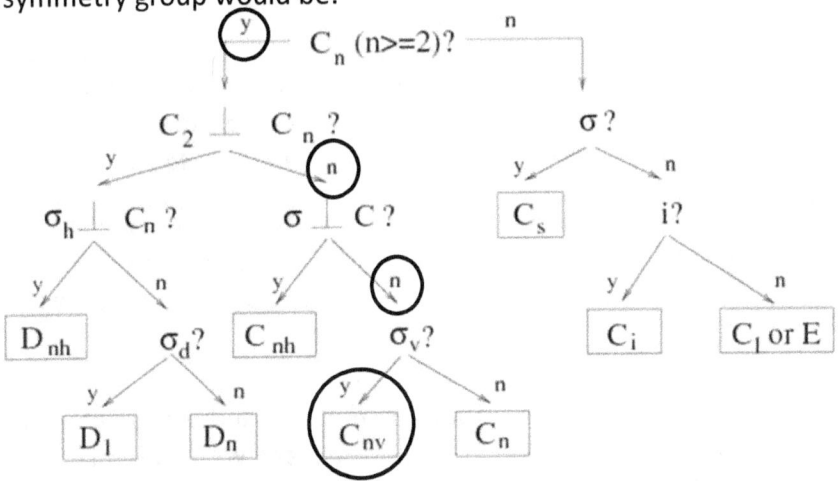

The character table of C2v shows us that there are 3 vibrations that are IR-active: **A1, B1 and B2**. Therefore we have to expect 3 strong peaks in the IR spectrum.

In the same way we can analyze all possible patterns for a carbonyl compound.

Character table for C_{2v} point group

	E	C_2 (z)	σ_v(xz)	σ_v(yz)	linear, rotations	quadratic
A_1	1	1	1	1	z	x^2, y^2, z^2
A_2	1	1	-1	-1	R_z	xy
B_1	1	-1	1	-1	x, R_y	xz
B_2	1	-1	-1	1	y, R_x	yz

Importance of ligand strength:

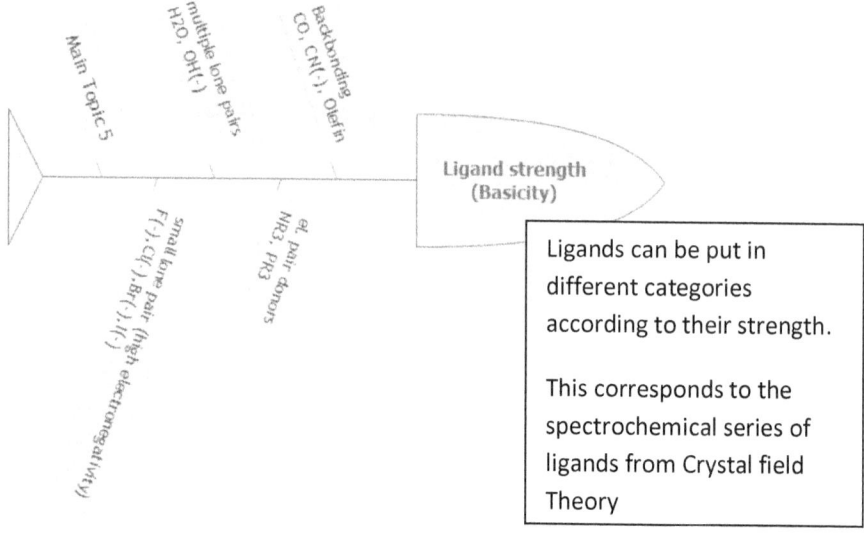

Ligand strength
(Basicity)

Ligands can be put in different categories according to their strength.

This corresponds to the spectrochemical series of ligands from Crystal field Theory

For industrial scale, using carbonyl complexes and UV-radiation are far too expensive; instead a complex is chosen with a metal that tolerates also 16 electrons (and therefore the loss of one ligand) – this is called <u>dissociative</u> mechanism:

$ML_n \rightarrow ML_{n-1} + L$ (slow)

(18 el.)   (16 el.)

$ML_{n-1} + X \rightarrow ML_{n-1}X$ (fast)

(16 el.)   (18 el.)

27	28
Co	Ni
Cobalt	Nickel
45	46
Rh	Pd
Rhodium	Palladium
77	78
Ir	Pt
Iridium	Platinum

The choice of a suitable solvent which can stabilize 16 electron species is essential for this mechanism. This is possible especially with Co and Ir complexes.

Which step in the following reactions is dissociative and which is associative?

3

$$- C_2H_4 \nearrow (C_2H_4)PdCl_2 \searrow CO$$

$$(C_2H_4)_2PdCl_2 \qquad\qquad ? \qquad\qquad (C_2H_4)(CO)PdCl_2$$

$$CO \searrow \qquad\qquad \nearrow - C_2H_4$$

$$(C_2H_4)_2(CO)PdCl_2$$

Write the total number of valence electrons for each of the four species.
Which reaction path would be comparable to an organic S_N1 and which to an S_N2 reaction ?

In a dissociative mechanism, is the rate of the ligand exchange dependent on the concentration of the new ligand X ?

4

Notice that the oxidation number of the metal does NOT change during the ligand exchange

In which direction would the reaction go according to what we know about ligand strength ?

$L_nM\text{-}CO + P(OMe)_3 \quad ==== \quad L_nM\text{-}P(OMe)_3 + CO$

5a

$(NH_3)_2PtCl_2 + H_2O \quad ==== \quad (NH_3)_2Pt(Cl)(H_2O)^{(+)} + Cl^{(-)}$

5b

$Ni(NH_3)_2(H_2O)_2 + H_2N\text{-}CH_2CH_2\text{-}NH_2 \quad === \quad Ni(NH_2\text{-}CH_2CH_2$

5c

(2) OXIDATIVE ADDITION

This reaction is especially useful, because it causes the break of a bond in a ligand molecule. This offers the chance to create a new bond to another organic molecule:

$$L_nM + \begin{array}{c} X \\ | \\ Y \end{array} \longrightarrow L_nM \begin{array}{c} X \\ \diagdown \\ Y \end{array}$$

Typical ligands are Hydrogen, alkyl and olefin compounds. In the case of a double bond, only the π-bond will break and we generate a new C-C single bond instead:

$$L_nM + X=Y \rightleftarrows L_nM \begin{array}{c} X \\ \diagup | \\ Y \end{array}$$

The <u>reverse reaction</u> is also possible and is called "reductive elimination" – in that case a new bond between X and Y is formed.

Which product would you expect from the following reaction ?

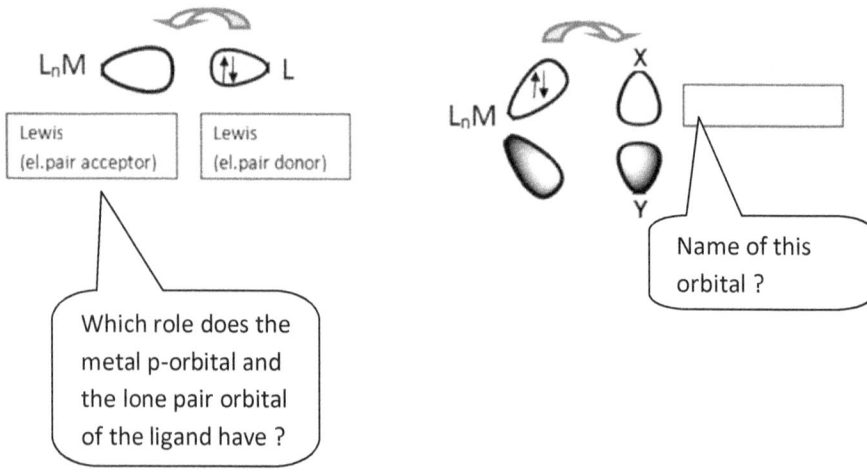

How can we explain this bond-splitting effect of the metal ?

L$_n$M ⬭ (↑↓) L

Lewis
(el.pair acceptor)

Lewis
(el.pair donor)

Which role does the metal p-orbital and the lone pair orbital of the ligand have ?

L$_n$M

Name of this orbital ?

Which products do we expect from following reactions ?
(draw the correct configuration)

$\boxed{7}$

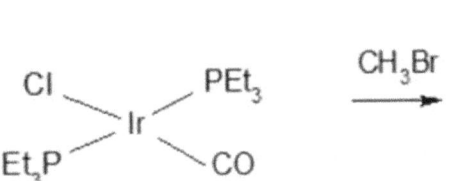

$$\xrightarrow{H_2}$$

$$\xrightarrow{CH_3Br}$$

Also indicate the
oxidation number of
the metal and the
number of valence
electrons.

$$\xrightarrow{HI}$$

Which of the following complexes can undergo oxidative
addition of methyl-chloride ?

$\boxed{8}$

$ZrCp_2(CH_3)(Cl)$ $Na_2[FeCO_4]$

$[RhI_2CO_2]^{(\cdot)}$ $Ir(PPh_3)_2(CO)(Cl)$

Remember - Oxidative Addition
means:
(a) the **oxidation number** of the
metal increases by +2
(b) the **number of valence**
electrons rises by +2

Please write the reaction equation, the oxidation number of the
metal and number of valence electrons.

Which of the following complexes is more reactive towards
oxidative addition of H_2 ? Explain your reason.

$Rh(PPh_3)_3(Cl)$ or $Rh(Ph_3)_2(CO)(Cl)$

$\boxed{9}$

$Ir(PPh_3)_2(CO)(Cl)$ or $Ir(PPh_3)(CO)(Cl)_2$

Which reaction do you expect between dimethyl-palladium(II)
and methylchloride ? Explain the reaction using simple MO
theory.

$\boxed{10}$

(3) REDUCTIVE ELIMINATION

The "reductive elimination" is just the reverse reaction of oxidative addition:

Two ligands on a metal combine to form a new (organic) molecule:

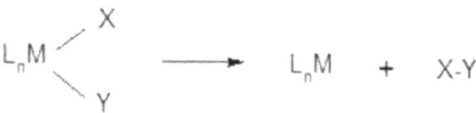

Write the reaction for the elimination of hydrogen in this molecule:

$$\text{(structure)} \longrightarrow \boxed{11}$$

Also indicate the oxidation number of Ir and the number of valence electrons.

An elimination (and therefore bond-formation) speed for hydrocarbons can be observed as follows:

$L_nM(H)(H) \rightleftharpoons L_nM + H_2$ — Reaction is fast and reversible

$L_nM(H)(R) \rightleftharpoons L_nM + R\text{-}H$ — Reaction is very fast and rarely reversible

$L_nM(R)(R') \rightleftharpoons L_nM + R\text{-}R'$ — Reaction is slow and rarely reversible

Which direction is an oxidative addition and which one a reductive elimination ?

12

$$\begin{array}{ccc} & & \\ & & \end{array}$$

Check your answer by writing the oxidation number of Pt on both sides.

Olefin bonding

When an olefin is coordinated to a transition metal, does the π bond become stronger or weaker ?

Example: Zeises Salt

Compare the double bond in ethylene with the bond in the coordinated molecule.

13

Molecules with π bonds are able to act as electron donor ligands using their π electrons (instead of a lone-pair). But in addition they offer empty π* orbitals into which the metal can transfer it's d-electrons (backbonding) .

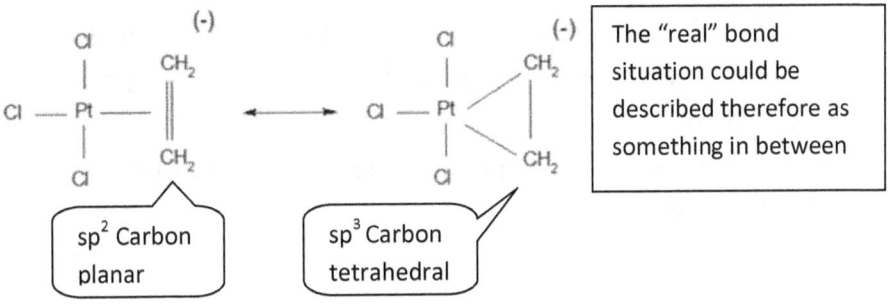

| sp² Carbon | sp³ Carbon | The "real" bond |
| planar | tetrahedral | situation could be |

sp² Carbon
planar

sp³ Carbon
tetrahedral

The "real" bond situation could be described therefore as something in between

(4) INSERTION / MIGRATION

This mechanism is about re-arrangement of bonds in a complex. The importance of this mechanism lies in the possibility to create new bonds between two ligands.

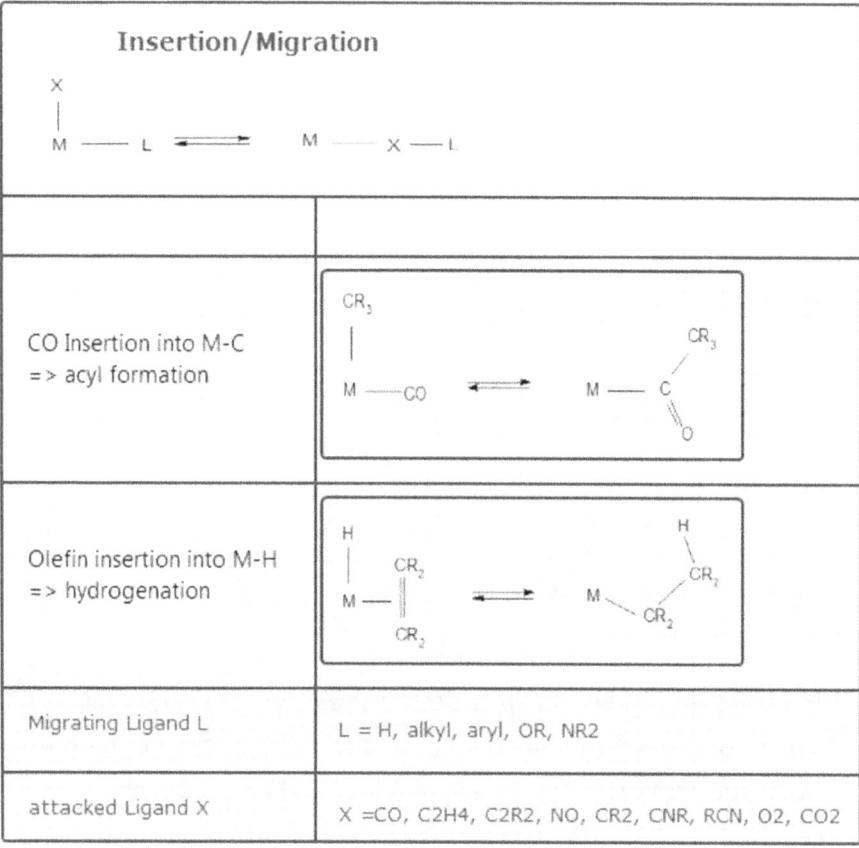

Insertion/Migration	
CO Insertion into M-C => acyl formation	
Olefin insertion into M-H => hydrogenation	
Migrating Ligand L	L = H, alkyl, aryl, OR, NR2
attacked Ligand X	X =CO, C2H4, C2R2, NO, CR2, CNR, RCN, O2, CO2

In fact the "insertion" for example of CO into a M-C bond is the wrong mechanism – it could be observed that it is a migration of the alkyl onto the C of CO.
Therefore the ligand X is "passiv" while L attacks X – like an electrophil addition.

The reverse reaction for olefins is called "β-H elimination" – an alkyl attached to a metal can lose one hydrogen in β-position, which then is directly bonded to the metal. The remaining alkyl then forms a double bond.

Example:

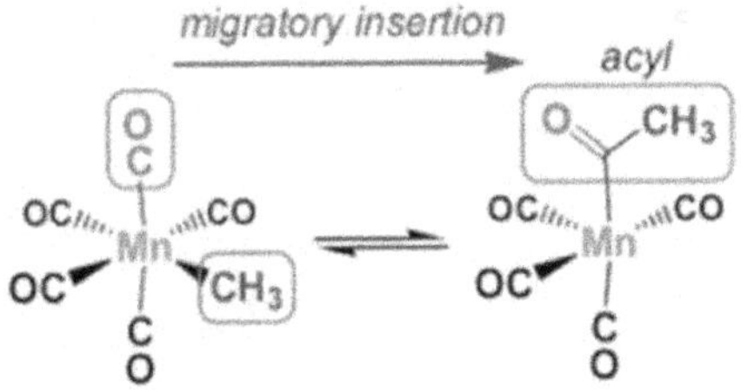

Write the number of valence electrons and oxidation number for both molecules.

14

Which reaction would you expect for this complex in the presence of a solvent that can act as ligand (ex. CH_3CN):

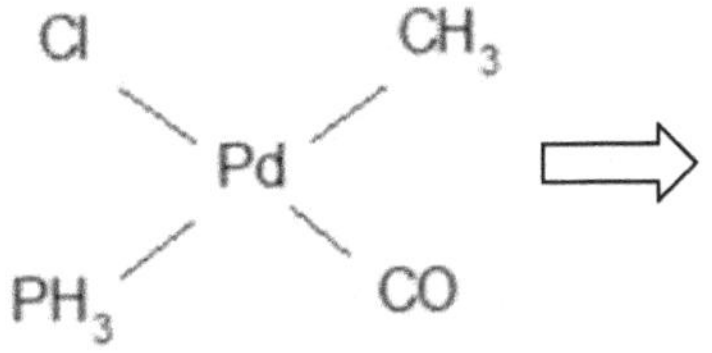

15

Describe the reaction path of the hydride migration onto a carbon of a close olefin ligand. Draw the intermediate and check oxidation number of Rh and number of valence electrons of the complex.

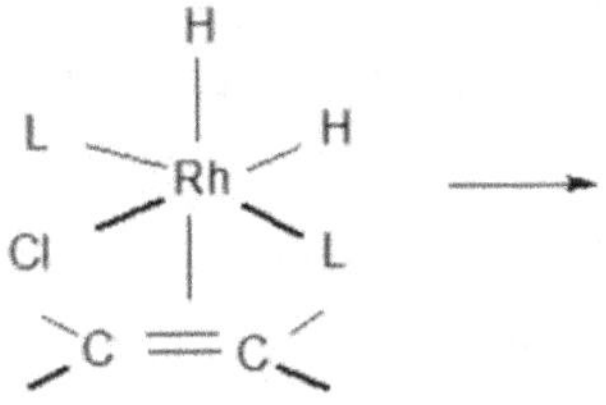

16

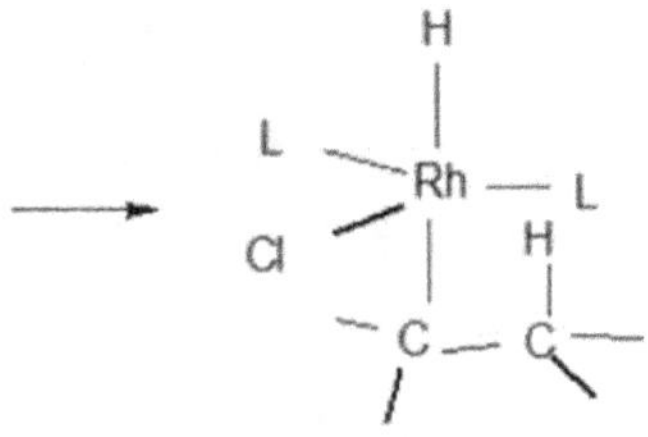

CATALYTIC CYCLES

A catalytic cycle is a process in (homogenous) catalysis, where a metal complex reacts with added molecules, transforming them, releasing a new molecule and be regenerated, so the whole process starts again:

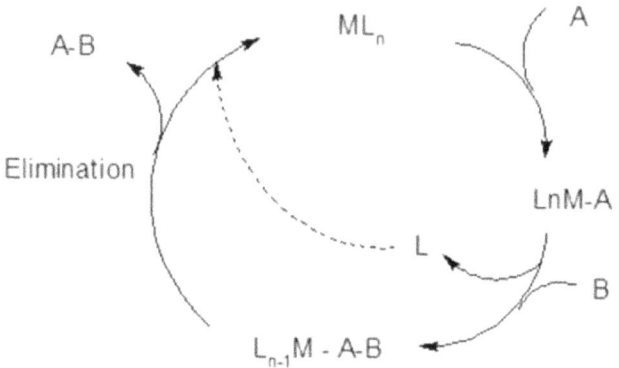

Rearrangement

This is a very generalized view: the essential idea is that a coordinatively unsaturated complex ML_n is able to add an external new ligand molecule and – by releasing one of its ligands L – a second external ligand. Through re-arrangements the two new ligands can be combined to be released as a new molecule and the original complex restored.

To achieve this, the metal M must have the ability to exist as stable 16- or even 14-electron species in order to accept 2 more electrons from an external ligand A.
Normally only second and third ow transition metals are "flexible" enough to perform such changes. The next section demonstrates some typical examples of catalytic cycles and you are invited to analyze (and understand !) each step in the sequence.

(a) Hydrogenation of olefins

Simplified mechanism:

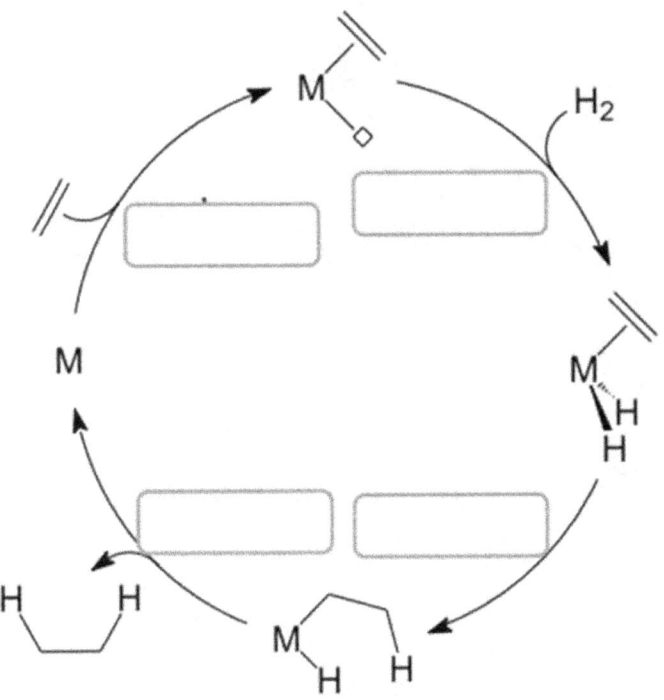

Indicate the names of the 4 steps. | 17 |

In case the metal M is Pd - give the oxidation number of the metal and total numer of valence electrons in each step.

| 18 |

Which kind of tranistion metal is most likely to perform such reactions ?
(early, middle or late transition metals) | 19 |

(b) Wilkinson's catalyst – olefin hydrogenation

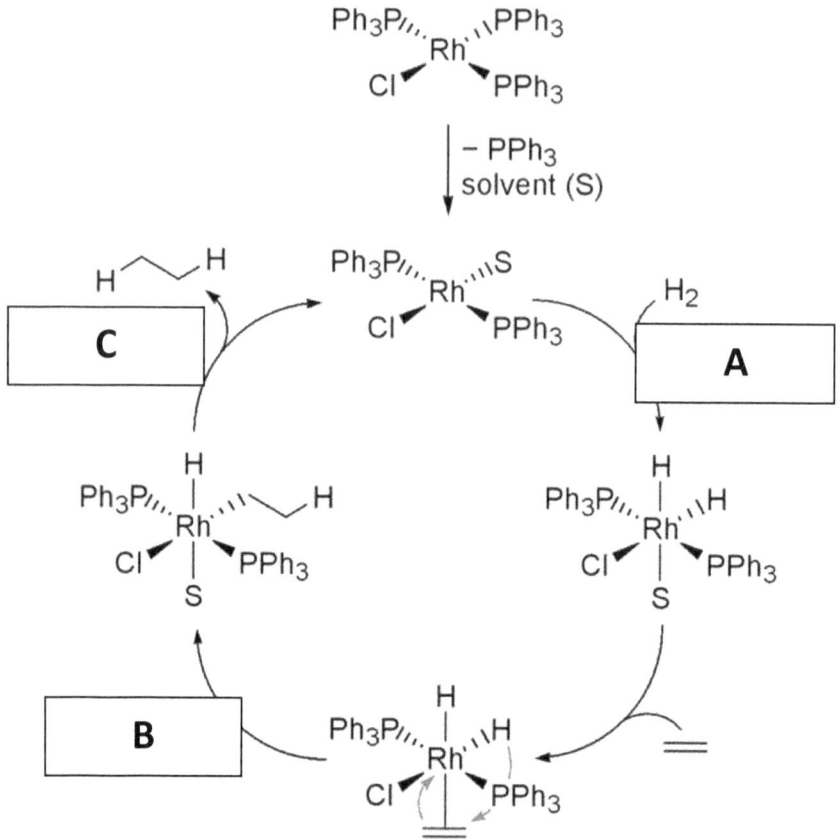

(from: https://commons.wikimedia.org/wiki/
File:Catalitic_cycle_for_hydrogenation_with_Wilkinson%27s_catalyst.png)

Assign to steps A - C what kind of basic step it represents.

20

Determine the oxidation number and valence electron count of each species - which species would be "satisfied" in it's state and which not ?

21

(c) Monsanto acetic acid process

(https://commons.wikimedia.org/wiki/File:Monsanto_process_catalytic_cycle.svg)

This process is more complex because 2 cycles work together:

- Methanol reacts with HI to give methyl-iodide,
- which can react with the Rhodium catalyst.
- Finally the catalyst transforms methyl-iodide + carbon monoxide to give
  acetyl iodide,
- which is hydrolyzed by water to yield the final acetic acid, thereby regenerating the HI

Name the steps 2- 5 and indicate the oxidation number of Rh and total number of valence electrons.

22

(d) Hydroformylation

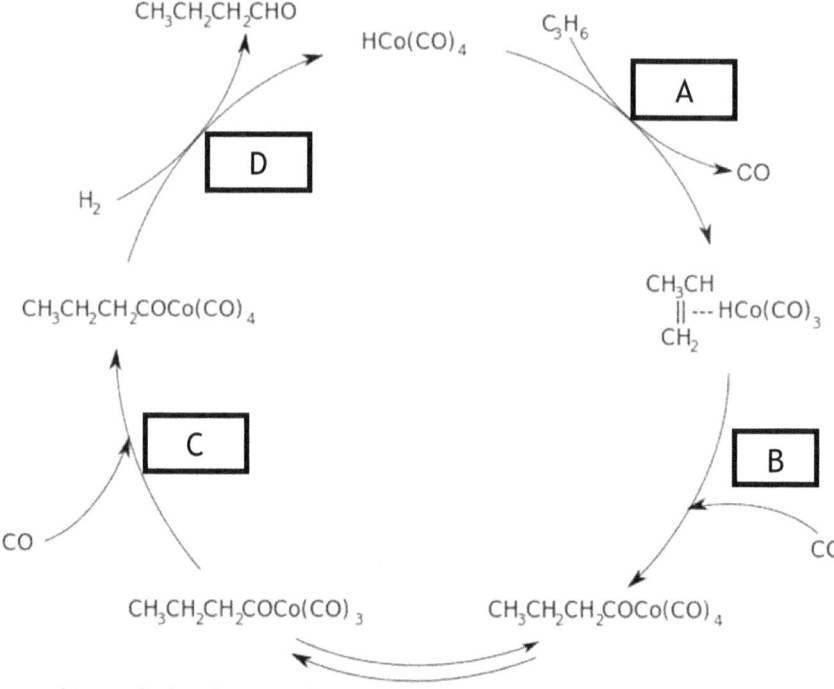

$CH_3CH_2CH_2CHO$

$HCo(CO)_4$

C_3H_6

A

D

CO

H_2

CH_3CH
$\quad \| \; --- HCo(CO)_3$
CH_2

$CH_3CH_2CH_2COCo(CO)_4$

C

B

CO

CO

$CH_3CH_2CH_2COCo(CO)_3$

$CH_3CH_2CH_2COCo(CO)_4$

(By Aushulz - Own work, CC BY-SA 3.0,
https://commons.wikimedia.org/w/index.php?curid=7129996)

Complete the summary formula for propylene transformation (without catalyst):

C_3H_6 +   --->   23

Write the structures of all Co-species and indicate the name of steps A - D, together with the oxidation number and valence electron count of Cobalt.

24

(e) Hiyama Coupling

$$R'-X \quad + \quad R-SiR''_3 \quad \xrightarrow[\text{Pd(OAc)}_2]{\text{TBAF}} \quad R-R'$$

R = Aryl   R' = Aryl
X = Br   SiR''$_3$ = SiMe$_3$

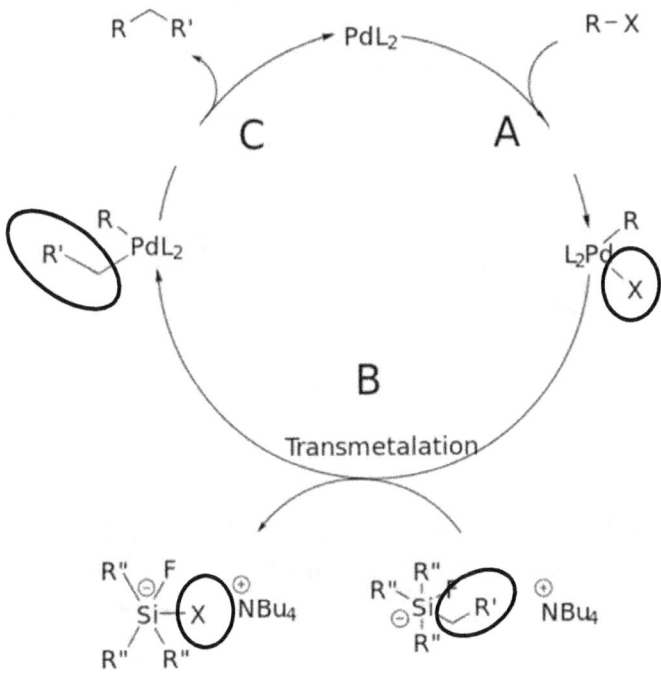

(By JABuonomo - Own work, CC BY-SA 3.0,
https://commons.wikimedia.org/w/index.php?curid=17610204)

Here is a new mechanistic step: "Transmetalation"

It is a transfer of ligands from one metal to another: In this case the ligand X is transferred to Silicium and the –CH2-R' alkyl ligand on Si to Pd.

Name the steps A and B, and the oxidation numbers of Pd and valence electron count for each species.

25

(f) Pd C-C coupling

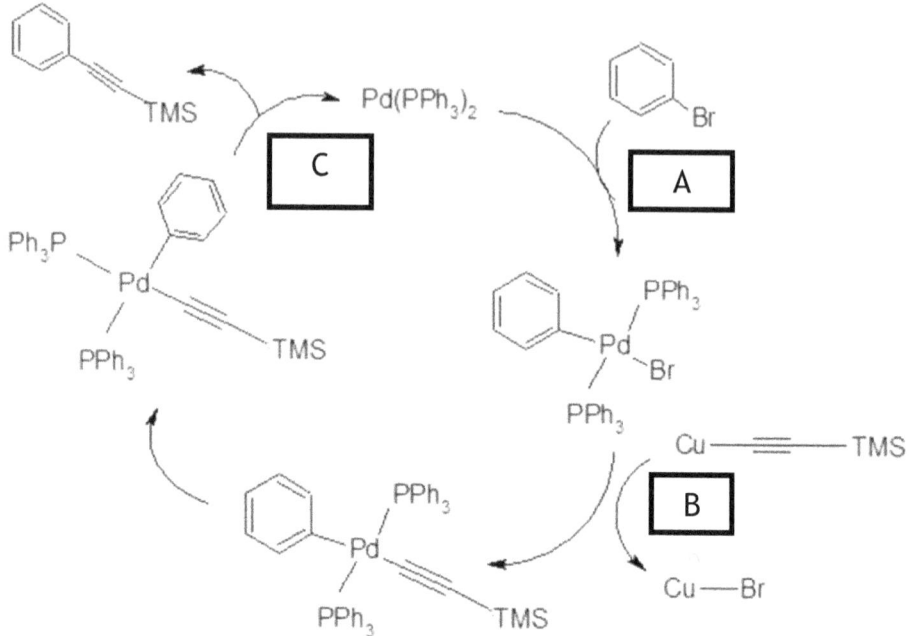

Please name the steps A- C and indicate the oxidation numbers and valence electron counts for the Pd-complex

26

(g) Heck Coupling

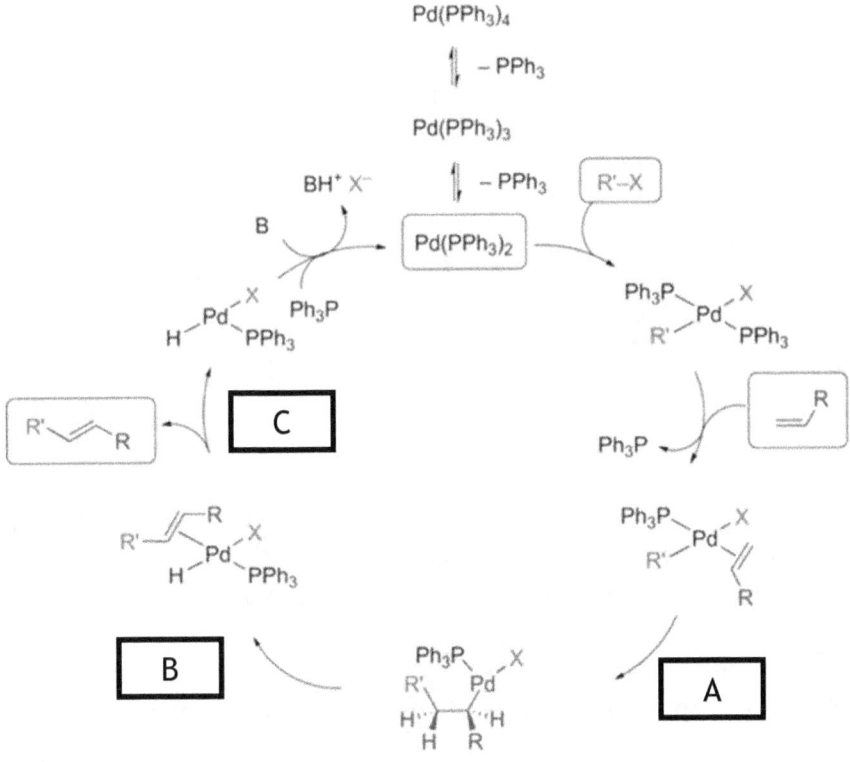

(By Mfomich - Own work, CC0,
https://commons.wikimedia.org/w/index.php?curid=27786436)

*Again: please name the steps A- C and indicate the oxidation
numbers and valence electron counts for the Pd-complex*

27

(h) Tsuji-Trost Cycle

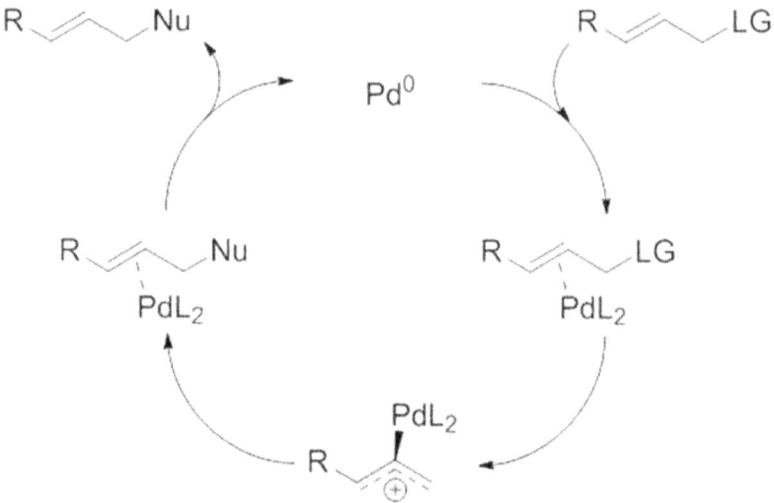

(By Mstreeter15 - Own work, CC BY-SA 3.0,
https://commons.wikimedia.org/w/index.php?curid=29970448)

LG = leaving group, for example halogen
Nu = nucleophil, for example hydride or carbenium-ion

The special mechanism here is the change is coordination of
the allyl-ligand, allowing the leaving group to go out and the
attack of a nucleophil on the allyl.

Explain how this cycle is different from the previous ones.

28

These examples covered only a fraction of the nearly endless possibilities to use the properties of transition metal compounds for catalysis. Many more examples can be found here:
http://www.ecompound.com/Reaction reference/reaction_reference_top.htm

By now you hopefully got a brief insight and catalysis lost its strangeness to you !

ANSWERS TO PROBLEMS

[1]				
Ligands:	3 PPh3: 6 el CO: 2 el. H(-): 2 el. -1	2 PMe3: 4 el. 3 CO: 6 el. =C: 4 el. -2	H: 2 el. -1 Cp: 5 el. -1 2 CO: 4 el. PR$_3$: 2 el.	Cp: 5 el. -1
Metal ox.no.	Rh +1 => 8 el.	Cr +2 => 4 el.	Mo +2 => 4 el.	Co +2 => 7 el.
Total:	18 el.	18. El.	17 el.	17 el.

[2]	Ph$_3$P- Pd -PPh$_3$			H-Co(CO)$_3$
Ligands:	PPh$_3$: 2 el.	Cp: 5 el. -1 CO: 2 el.	Cl: 2 el. -1 CO: 2 el. PPh$_3$: 2 el.	H: 2 el. -1 CO: 2 el.
Metal ox.no.	Pd 0 => 10 el.	Mn +1 => 6 el.	Ir +1 => 8 el.	Co +1 => 8 el.
Total:	14 el.	17 el.	16 el.	16 el.

[3]

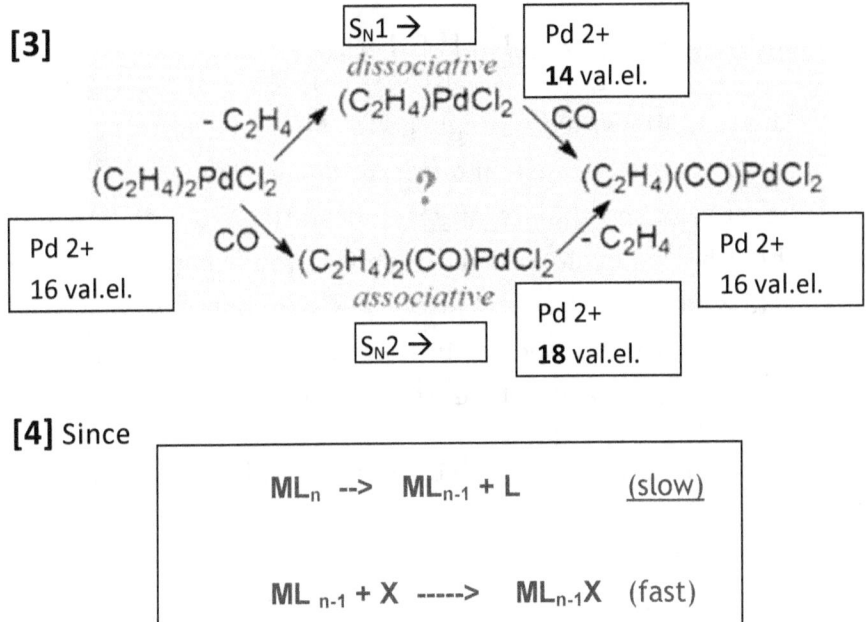

[4] Since

$$ML_n \;\; \text{-->} \;\; ML_{n-1} + L \qquad \text{(slow)}$$

$$ML_{n-1} + X \;\; \text{----->} \;\; ML_{n-1}X \;\; \text{(fast)}$$

the concentration of X does <u>not</u> affect the speed of the overall reaction.

[5] $L_nM\text{-}CO + P(OMe)_3 \;\; \Longleftarrow \;\; L_nM\text{-}P(OMe)_3 + CO$

The reaction should go towards the left, because CO is a stronger ligand than Phospite (But since CO is a gas, it can be easily taken out of the equation. Under this condition, the reaction would go to the right)

$(NH_3)_2PtCl_2 + H_2O \;\; \Longrightarrow \;\; (NH_3)_2Pt(Cl)(H_2O) \; (+) + Cl(-)$

This reaction would go to the right, water is a stronger ligand than chloride.

(Nevertheless the difference in ligand strength is not big, we would find a mix of both species in the equilibrium)

$Ni(NH_3)_2(H_2O)_2$ + $H_2N-CH_2CH_2-NH_2$ =>>
$Ni(NH_2-CH_2CH_2-NH_2)$ $(H_2O)_2$ + 2 NH_3

The reaction would strongly go to the right – in this case the reason is not because of different ligand strength but because of the chelate-effect - once attached with one end, the other end will quickly attach too, replacing a single-attached ligand. You can also argue with entropy: on the left side we have two species but on the right side three, therefore increasing the entropy significantly.

[6] We could expect a mixture of cis- and trans complex:

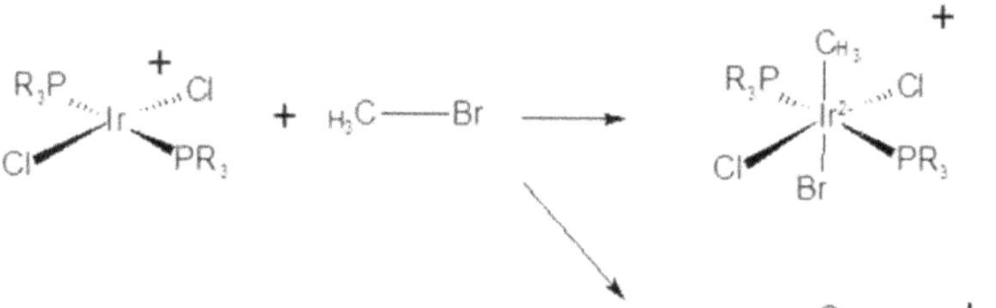

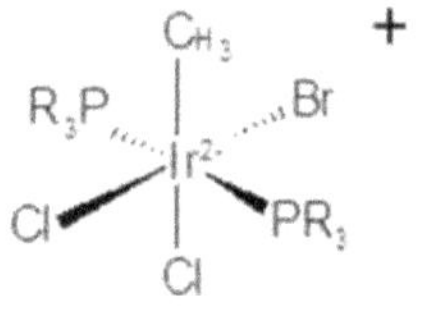

Two possible mechanisms would be:

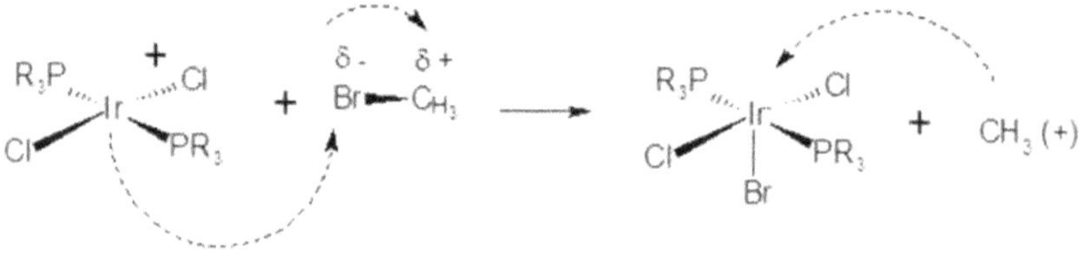

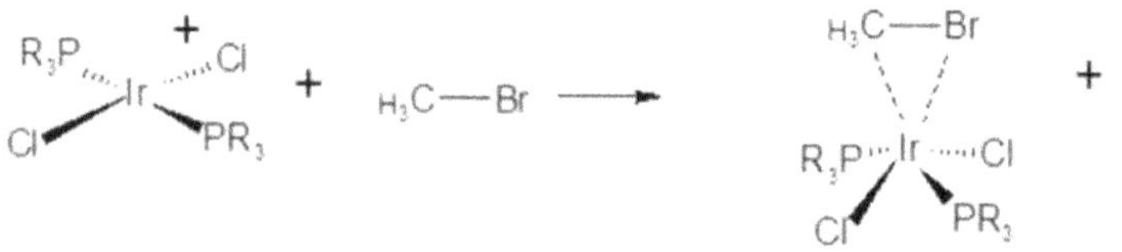

It would depend also on the solvent – a polar solvent would promote a kind of SN1 process, otherwise a SN2-like process could occur, leading to the cis-product.

[7] Oxidative Additions:

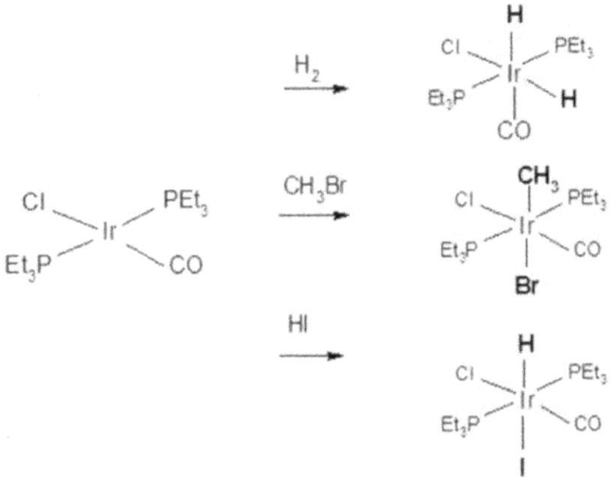

We can assume that polar molecules like methylbromide and hydrogeniodide would add to the complex in a SN1 manner (s. question 6), but non-polar molecules like hydrogen would first add to the metal as a whole, then splitted by d-electrons of the metal.

[8] Oxidative addition of:

$ZrCp_2(CH_3)(Cl)$ (16 el, tetrahedral, Zr ox.no. +4)
is not possible, since Zr is already in its maximum oxidation state (+4)

$Na_2[FeCO_4]$ (18 el., tetrahedral, Fe ox.no. -2)
No reaction, because $[Fe(CO)_4]^{2-}$ has already 18 electrons and the addition of CH_3-I would give a 20 electron complex.

$[RhI_2(CO)_2]$ (-) (16 el., square planar, Rh ox.no. +1)
is possible since Rh has an ox.no. of +1 and 16 el., which can be easily increased to +3 and 18 el.:

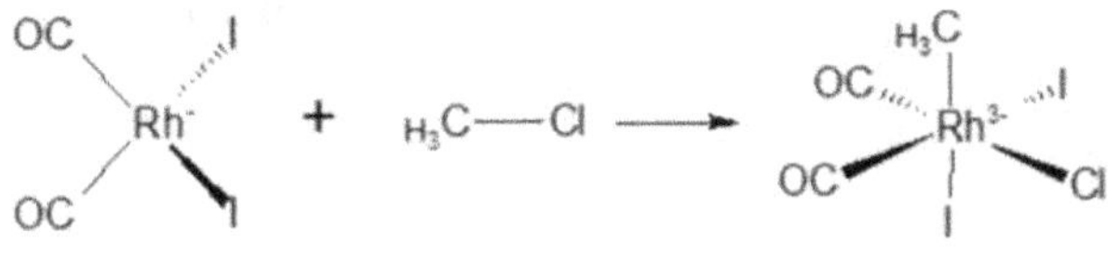

Ir(PPh$_3$)$_2$(CO)(Cl) (16 el., square planar, Ir ox.no. +1)
Similar case: Ir can be oxidized to +3 and the valence electron number from 16 to 18:

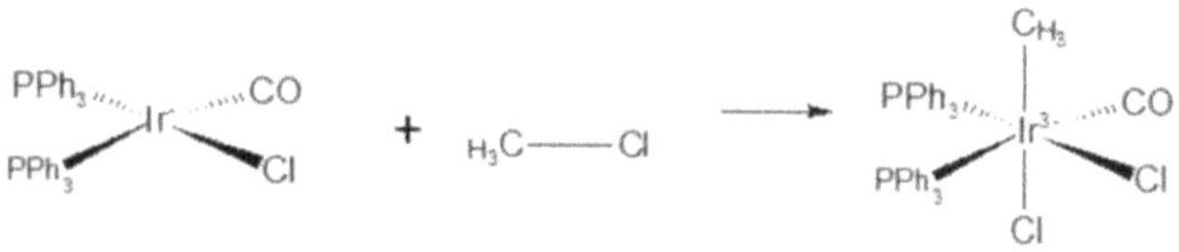

[9] Higher susceptibility towards oxidative addition:

Rh(PPh$_3$)$_3$(Cl) or   Rh(Ph$_3$)$_2$(CO)(Cl)
CO will pull electron density away from the central metal, therefore the complex will be less able to perform oxidative addition than with the phosphine ligands.
Therefore Rh(PPh$_3$)$_3$(Cl) should be more reactive.

Ir(PPh$_3$)$_2$(CO)(Cl) or Ir(PPh$_3$)(CO)(Cl)$_2$
Similar case: an additional chloride ligand is likely to reduce the electron density of the metal, therefore Ir(PPh$_3$)$_2$(CO)(Cl) should be more reactive.

[10] Reaction between Pd(Me)$_2$ and Me-Cl:

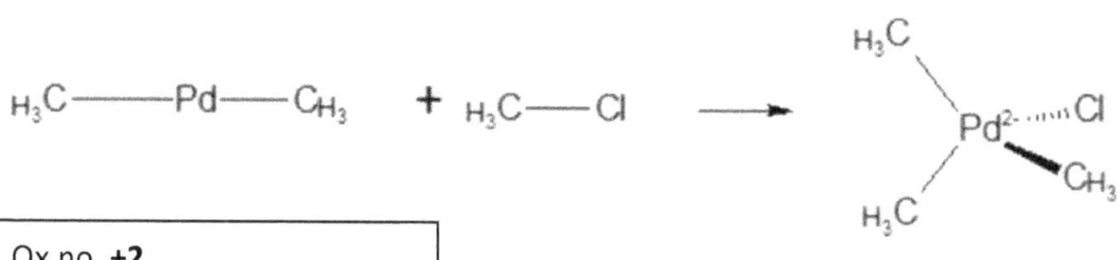

Ox.no. +2
Valence electrons:
8 (Pd2+) + 4 (2x CH3) = 12 el.

Ox.no. +4
Valence electrons:
6 (Pd4+) + 6 (3x CH3) + 2 (Cl) = 14 el.

Pd metal can interact with the C-Cl bond in 2 ways:

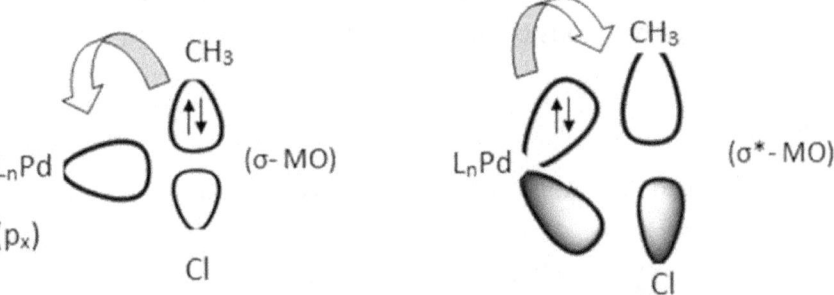

The bonding C-Cl σ MO can bind to the empty px orbital of the metal	The (filled) dxz Orbital of the metal can put electrons into the C-Cl σ* MO

Due to the second, "backbonding" effect, the C-Cl bond can be split and two new σ-bonds between metal and CH3 and Cl can be formed:

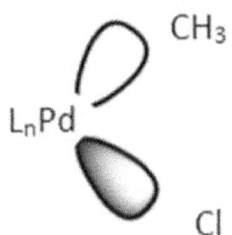

This reaction would leave the methyl and chlorine group in a cis-position to each other.

[11] Reductive elimination of hydrogen:

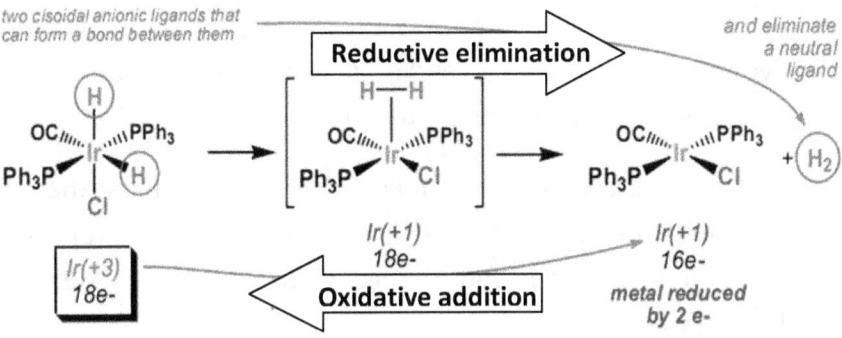

(http://www.umich.edu/~chemh215/W10HTML/SSG3/ssg5/leading.html)

This reaction is common for H-M-H to form hydrogen and for R-M-H to form a R-H molecule. In rare cases is also possible that two carbon ligands combine together to form a new C-C bond.

[12] Oxidative Addition and reductive elimination:

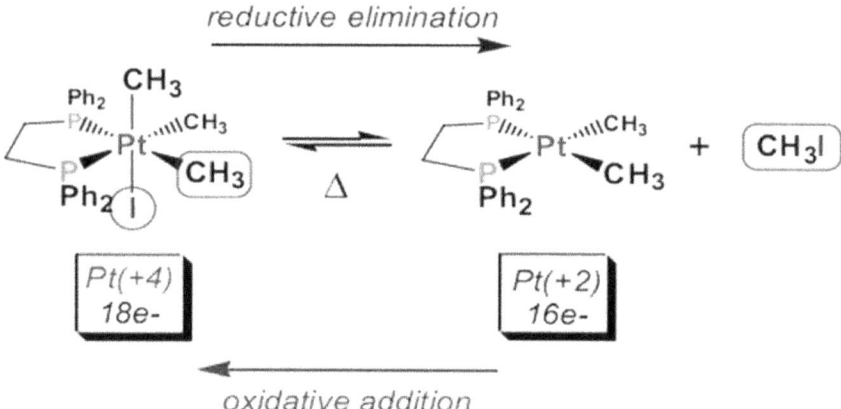

In this example, under heating, the oxidative addition is preferred.

[13] The ethylene bond in free ethene and bond to a transition metal:

A π-bond is symmetric for rotation and therefore has the same symmetry as a metal s, p_z and d_{z2} orbital: electron density from the π -bond will go to the metal orbitals forming a bond.

BUT: there is also an anti-bonding π^* orbital which has the same symmetry as a d_{xz} metal orbital, that means that electron density from the metal can go into this ligand orbital and makes the π -bond <u>weaker</u>.

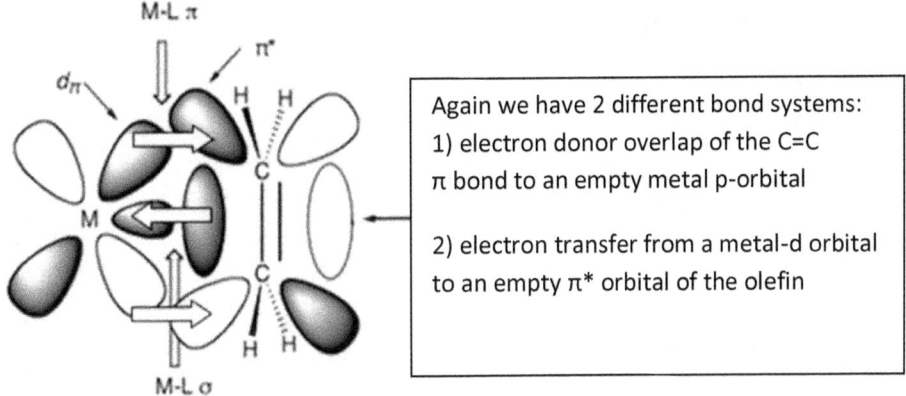

Again we have 2 different bond systems:
1) electron donor overlap of the C=C
π bond to an empty metal p-orbital

2) electron transfer from a metal-d orbital
to an empty π* orbital of the olefin

[14] Insertion/Migration:

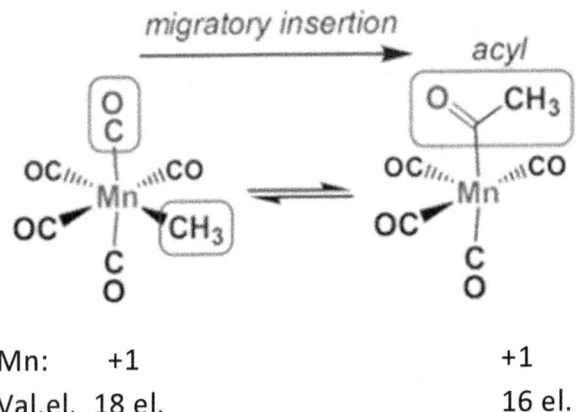

Mn: +1   +1
Val.el. 18 el.   16 el.

The oxidation number of the metal does not change, but due to the formal loss of one ligand, the valence electron count is two less.

[15] Insertion/Migration:

Pd:	+2		+2
Val.el.	16		16

A carbonyl-ligand in proximity to an alkyl-ligand often leads to a so-called "CO-insertion" into the metal-alkyl bond. In fact it is a <u>migration</u> of the alkyl-group onto the carbon of the CO-ligand. This is especially likely if a solvent (S) is present that can compensate for the "loss" of 2 electrons when the alkyl-ligand is migrating to CO.

[16] Olefin-"insertion" / Hydrogen-migration

Rh:	+3		+3
Val.el.	18		16

The olefin "shifts" towards the hydrogen-ligand and forms a kind of 4-ring in an intermediate state.
As in the previous question, the migration is preferred when a solvent is present to bring the electron count to 18 gain.

[17] [18] Catalytic cycle – olefin oxidation

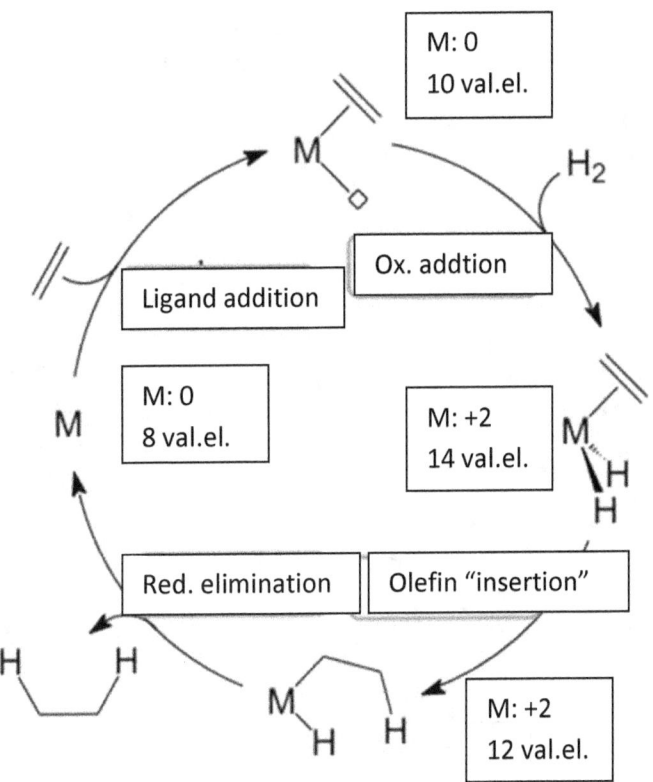

[19] Most likely metals to perform this kind of reactions are the late transition metals, in the second and third row.

Many of these metal complexes can exist with 12 – 18 valence electrons, they form square-planar 4-coordinate species which can easily acquire two more ligands by oxidative addition. The 2nd and 3rd row metals also have more space around the center for ligand re-arrangements.

[20] [21] Wilkinson catalytic cycle

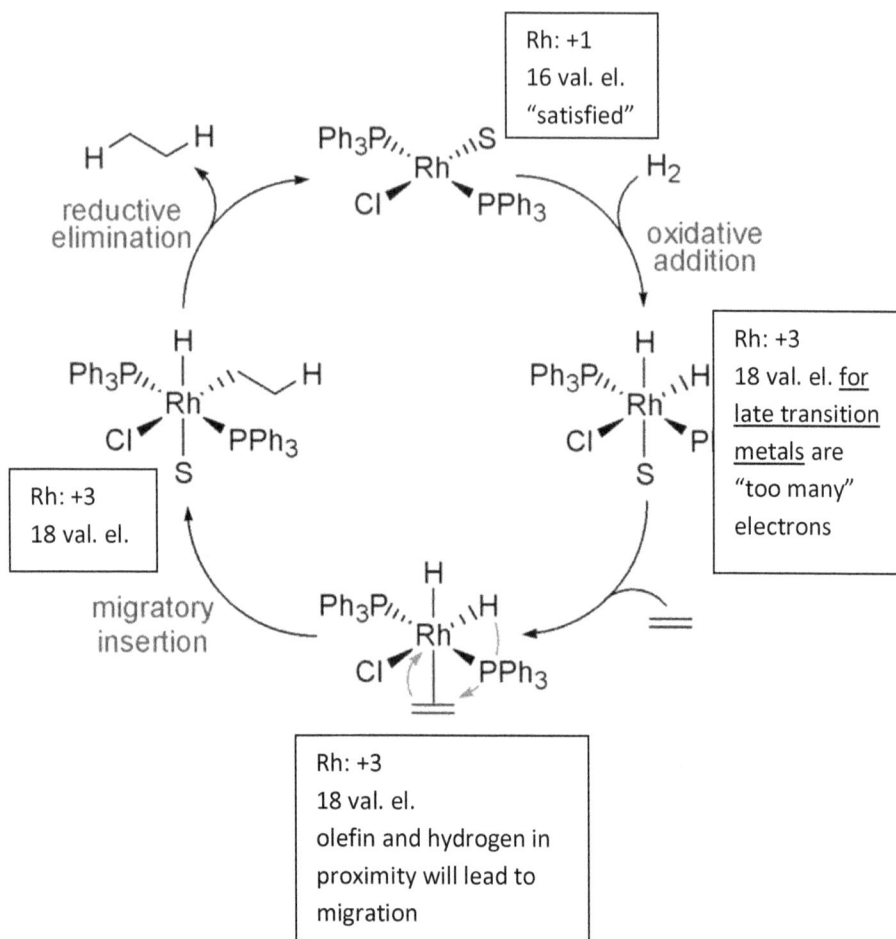

Rh: +1
16 val. el.
"satisfied"

oxidative
addition

Rh: +3
18 val. el. <u>for</u>
<u>late transition</u>
<u>metals</u> are
"too many"
electrons

reductive
elimination

Rh: +3
18 val. el.

migratory
insertion

Rh: +3
18 val. el.
olefin and hydrogen in
proximity will lead to
migration

[22] Monsanto Acetic acid process

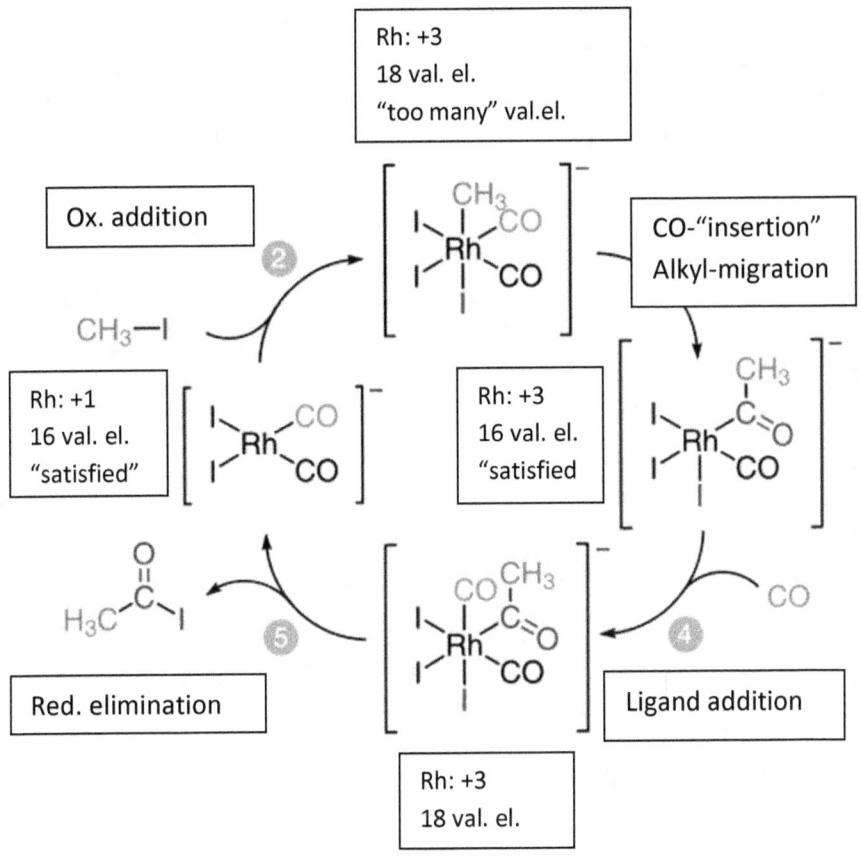

37

[23] [24] Hydroformylation: $C_3H_6 + CO + H_2 \rightarrow CH_3CH_2CH_2CHO$

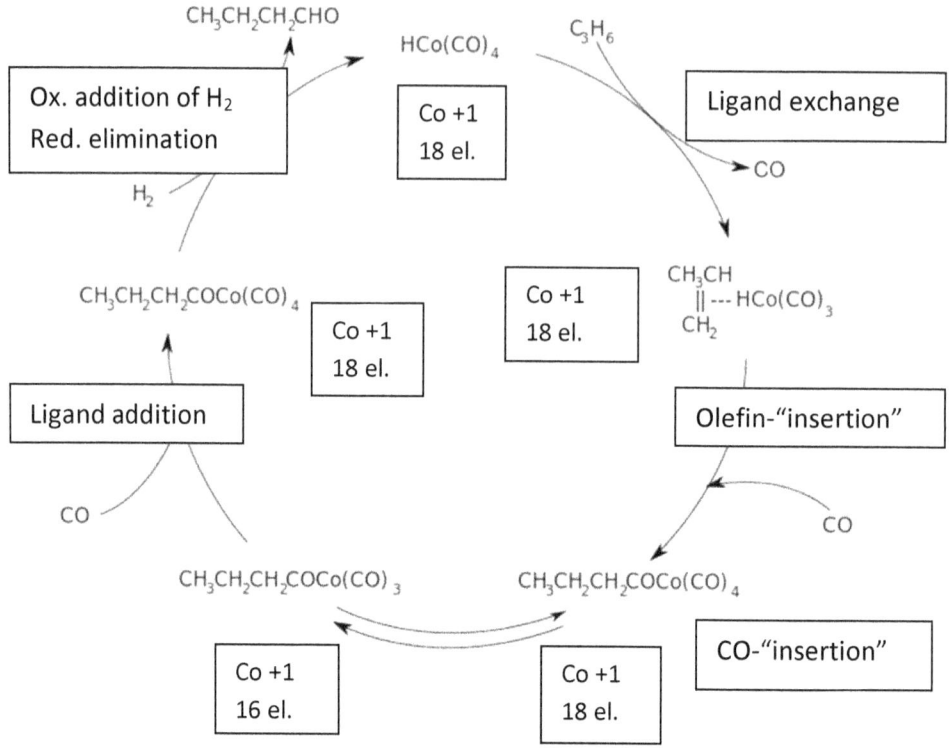

Detailed mechanism:

[25] Hiyama Coupling

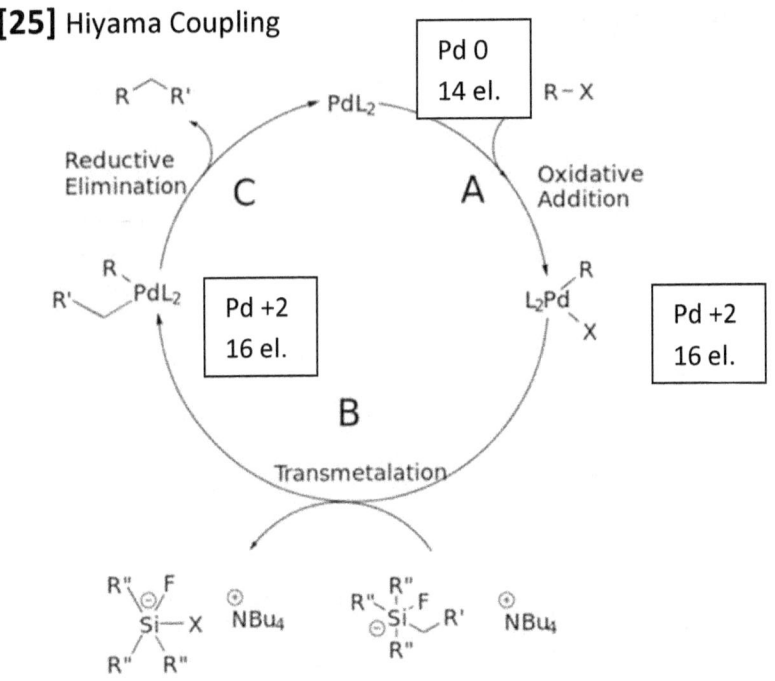

[26] Pd C-C coupling

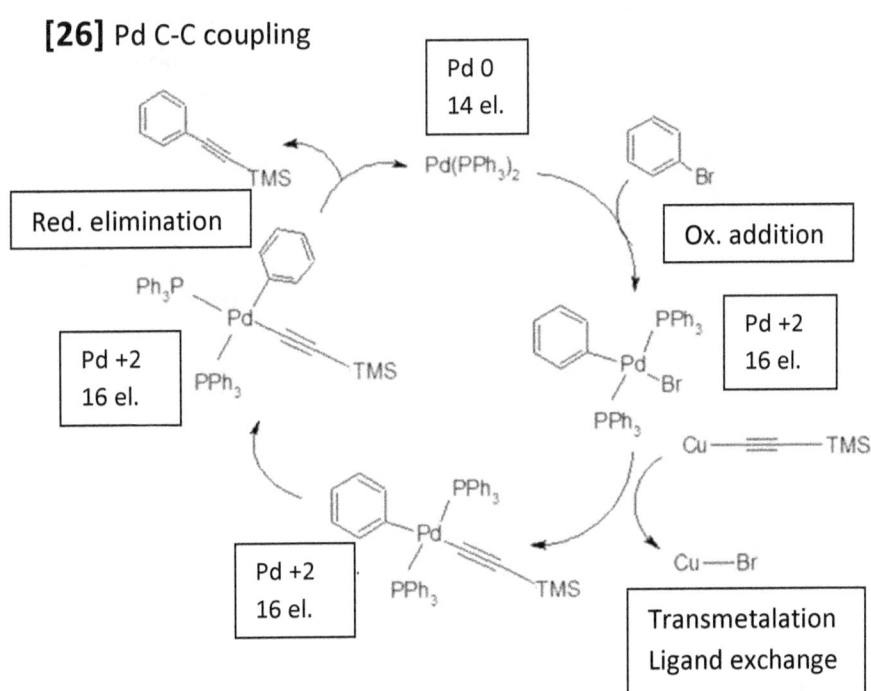

[27] Heck Coupling

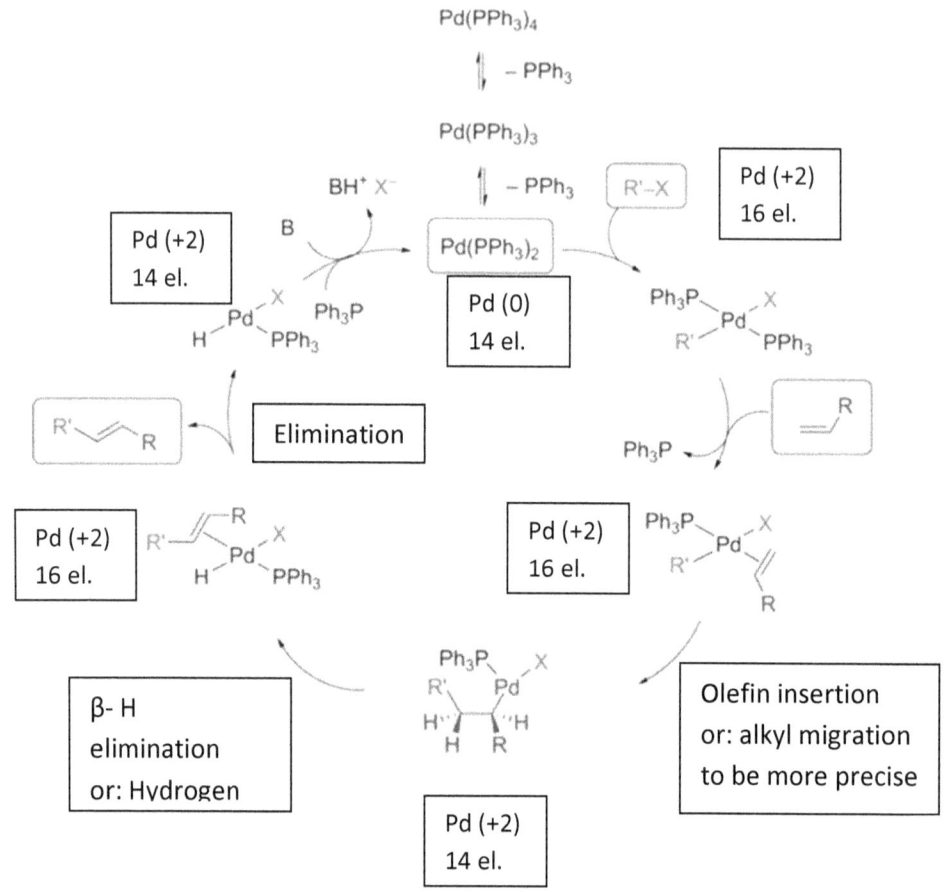

Pd(PPh₃)₄

‖ – PPh₃

Pd(PPh₃)₃

‖ – PPh₃

R'–X

Pd (+2)
16 el.

BH⁺ X⁻

B

Pd(PPh₃)₂

Pd (0)
14 el.

Pd (+2)
14 el.

Pd with X, Ph₃P, H, PPh₃

Ph₃P Pd X
R' PPh₃

R'⌒R

Elimination

Ph₃P

⫽R (olefin)

Pd (+2)
16 el.

R'—R Pd X
H PPh₃

Pd (+2)
16 el.

Ph₃P Pd X
R' ⫽ R

β- H
elimination
or: Hydrogen

Ph₃P Pd X
R'
H H
H R

Olefin insertion
or: alkyl migration
to be more precise

Pd (+2)
14 el.

[28] Tsuji-Trost Cycle

This cycle is different from the previous examples by:

- shift of an olefin to transform into an allyl-ligand
- nucleophilic attack on a ligand from an external molecule
- loss of a ligand without reductive elimination

ABOUT THE AUTHOR

I was born in 1958 in Germany and studied Chemistry at the University of Constance. For my dissertation in theoretical and experimental organometallic chemistry I needed 5 years and then followed my PhD father to Zurich, Switzerland in 1989.

Changing to industrial chemistry at Ciba-Geigy, Sandoz and Novartis until 2000, I decided to move to the, then new, field of internet programming and worked in this area for several years.

Finally I came back to teaching chemistry at universities in Thailand with the focus on getting understanding of organometallic reactions and motivate to explore this fascinating world even more.

Right now I work as lecturer and researcher at the University of Phayao, Thailand. My main field is the exploration of transition metal catalysts for the promotion of important steps for organic synthesis.

www.ingramcontent.com/pod-product-compliance
Lightning Source LLC
Chambersburg PA
CBHW070413190526
45169CB00003B/1238